LAN
OUT
SERIES

Work Out

Chemistry

'A' Level

The titles
in this
series

MACMILLAN
WORK OUT
SERIES

Work Out

Chemistry

'A' Level

D. A. Burgess

MACMILLAN
EDUCATION

First published 1987
Reprinted 1988

Published by
MACMILLAN EDUCATION LTD
Houndmills, Basingstoke, Hampshire RG21 2XS
and London
Companies and representatives
throughout the world

Typeset by TecSet Ltd, Wallington, Surrey
Printed at The Bath Press, Avon

British Library Cataloguing in Publication Data
Burgess, D. A.
Work out chemistry 'A' level.—(Macmillan
work out series)
1. Chemistry—Examinations, questions,
etc.
i. Title
540'.76   QD42
ISBN 0–333–39766–5

To my students, from whom I never cease to learn

Contents

General Introduction

How to Use the Book

This book is intended for those studying for GCE 'A' level or Scottish 'H' Grade examinations. It is not a substitute for a textbook but aims to give a concise summary of a topic, followed by worked examples and further questions for the student to attempt.

The book could be used during a course of study to consolidate each topic as it is completed; at the end of a course as a major revision aid, or of course, both.

The summaries are intended to be sufficiently comprehensive to be useful as a quick reference source for questions other than those in the book. However, in some cases the worked examples give added information which it would have been tedious to repeat in the summary.

Questions have been selected from a variety of Examination Boards and include objective questions, structured questions and longer 'essay' questions. The worked examples do not give a model answer in the sense that nothing else will do; many questions may be treated in several ways, all equally acceptable. The real purpose is to help you to understand what an examiner is looking for in an answer. To this end, the author's comments are printed in **bold type**, to distinguish them from the suggested answer itself.

Objective questions come after the longer essay and structured questions simply because teaching points can more usefully be made in the longer questions in order to clear up difficulties and misconceptions. Later chapters often include a question which revises the work of earlier chapters, to show the links between topics. The great majority of questions in the book are of direct relevance to most syllabuses. Further questions with a star (*) have an answer in Section 16.2.

Questions that have deliberately been avoided because of limitations of space include (a) industrial processes, (b) special options such as biochemistry, metallurgy and colour chemistry, and (c) descriptive accounts of practical work.

Revision

(a) Obtain a syllabus and past examination papers. These may be bought from the Examination Boards listed on page xi.

(b) Keep a small booklet in which definitions, laws and principles may be stored. Make constant reference to it so that it becomes part of your 'armoury'. If your Board does not supply a Periodic Table for use in the examination, include in the booklet the first three rows (beginning with hydrogen, lithium and sodium, respectively). You should also add the first transition row. Persist until this upper part of the table can be written out from memory.

(c) Select the most important 30 or so organic reactions and put each one on a small 'flip' card (about 8 cm × 4 cm), using both sides of the card somewhat as follows:

side 1: $$A + B \xrightarrow{?} ? + D$$

side 2: $$A + ? \xrightarrow{H^+} C + D$$

Side 2 gives the answer to the queries on side 1 and vice versa. You thus have 60 or so questions with immediate answers, to help in learning factual material. The idea can be extended to other parts of the syllabus, of course, but don't overdo it. The pack of cards should be small enough to be carried in a pocket, for use in odd moments, such as waiting for a bus or train.

(d) Finally, *organise* your revision time. Use regularly certain evenings or days of the week and, if you are studying for two or more examinations, have a time-table. For example, Monday evening 7–8, chemistry; 8–8.30, break; 8.30–9.30, biology. No matter how interested or absorbed you become in chemistry(!), *stop* at 8 p.m. Have a timed break: put on a tape or record, have a coke, or whatever. Then resume with a different subject.

You must find your own best study hours; some people work best early in the morning, others in the evening. If you do not plan your revision systematically, it is quite astonishing how the weeks fly past with those ever-present good intentions never actually being implemented.

The Examination

(a) When you sit down, make sure your desk is firm and does not wobble.

(b) Don't clutter your desk with materials which are not likely to be needed. Do not use correcting fluid — mistakes should be crossed out lightly; they may in fact get you some marks.

(c) When you have a choice of questions, read the whole paper carefully first. Apart from the rather obvious fact that the questions you choose first are those that you can answer well, there is evidence that the brain operates at an unconscious level so that a question which at first seems difficult or obscure may become much plainer an hour or so later.

Do not be mesmerised by what you cannot (or think you cannot) do. Remember, if you are able to answer correctly only 50% of each of the required number of questions, you will be successful.

(d) Once again, keep a fairly strict timetable. Allocate a suitable time for each question (this can often be worked out from past papers beforehand). If it helps, write approximate times for changing over on the question paper.

(e) Try to classify each question; once you have realised the nature of the beast, you can decide whether to avoid it or tame it! This is where plenty of practice on past examination questions is of enormous help.

Don't be afraid (to change our metaphor) to abandon ship. If, after, say 15 or 20 minutes, you realise you are in a hopeless muddle, make a firm decision to leave it and try another.

(f) Finally, in setting down your answer to an essay-type ('free response') question, do follow the numbering of each part *as given in the question*. Do not use (1), (2), (3) if the question is subdivided (a), (b), (c).

It is the author's hope that this book will help you to have the success that you want.

London W7, 1986   D. A. B.

Acknowledgements

The author and publishers wish to thank the following who have kindly given permission for the use of copyright material:

The Associated Examining Board, Northern Ireland Schools Examination Council, Scottish Examination Board, Southern Universities' Joint Board, University of London School Examinations Board and University of Oxford Delegacy of Local Examinations for questions from past examination papers.

The University of London Entrance and Schools Examination Council accepts no responsibility whatsoever for the accuracy or method in the answers given in this book to actual questions set by the London Board.

Acknowledgement is made to the Southern Universities' Joint Board for School Examinations for permission to use questions taken from their past papers but the Board is in no way responsible for answers that may be provided and they are solely the responsibility of the author.

The Associated Examining Board, the University of Oxford Delegacy of Local Examinations, the Northern Ireland Schools Examination Council and the Scottish Examination Board wish to point out that worked examples included in the text are entirely the responsibility of the author and have neither been provided nor approved by the Board.

Every effort has been made to trace all the copyright holders but if any has been inadvertently overlooked the publishers will be pleased to make the necessary arrangements at the first opportunity.

Names and symbols used are in general those recommended by the Association for Science Education Report, *Chemical Nomenclature, Symbols and Terminology for Use in School Science*, 3rd edition. Numerical data were compiled from the *Chemistry Data Book*, by J. G. Stark and H. G. Wallace (1984) and the *Revised Nuffield Advanced Science Book of Data* (1984).

The author is indebted to various colleagues for constructive suggestions, in particular Dr P. Judd and Dr S. Wilson. Errors and misconceptions, of course, remain the author's responsibility.

Examination Boards for Advanced level

Syllabuses and past examination papers can be obtained from:

The Associated Examining Board (AEB)
Stag Hill House
Guildford
Surrey   GU2 5XJ

University of Cambridge Local Examinations Syndicate (UCLES)
Syndicate Buildings
Hills Road
Cambridge   CB1 2EU

Joint Matriculation Board (JMB)
78 Park Road
Altrincham
Cheshire   WA14 5QQ

University of London School Examinations Board (L)
University of London Publications Office
52 Gordon Square
London   WC1E 6EE

University of Oxford (OLE)
Delegacy of Local Examinations
Ewert Place
Summertown
Oxford   OX2 7BZ

Oxford and Cambridge Schools Examination Board (O&C)
10 Trumpington Street
Cambridge   CB2 1QB

Scottish Examination Board (SEB)
Robert Gibson & Sons (Glasgow) Ltd
17 Fitzroy Place
Glasgow   G3 7SF

Southern Universities' Joint Board (SUJB)
Cotham Road
Bristol   BS6 6DD

Welsh Joint Education Committee (WJEC)
245 Western Avenue
Cardiff   CF5 2YX

Northern Ireland Schools Examination Council (NISEC)
Examinations Office
Beechill House
Beechill Road
Belfast BT8 4RS

1 Matter, Masses and Moles

1.1 Introduction

This chapter covers some of the more elementary calculations in A-level chemistry. Underlying most of them is the chemist's unit of 'amount of substance', the mole. It is vital to make a determined effort to master this concept thoroughly, and to see the point of doing so, before going on to the remainder of the book.

1.2 Units of Matter

All forms of matter, whether solids, liquids or gases, may be classified as follows.

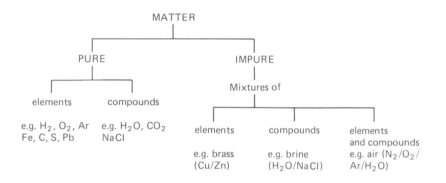

- **Compounds** can be resolved into elements.
- **Elements** can be resolved into atoms.
- **Atoms** can be resolved into protons, neutrons and electrons.

The **Atomic Number** (Z) of a species of atom (i.e. a nuclide) is the number of protons in the nucleus.

The **Mass Number** (A) of a nuclide is the sum of protons and neutrons in the nucleus.

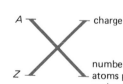

A nuclide, X, may be represented as shown on the left. An element can have several **isotopes**; these have the same value for Z but different values of A (different numbers of neutrons). There are about 1500 naturally occurring nuclides, comprising sub-sets of isotopes for most of the elements (see Fig. 1.1). Some elements appear to have only one isotope.

The **Relative Atomic Mass** (A_r) of an element is the weighted average of the mass numbers of the isotopes present (assumed constant). For example, chlorine

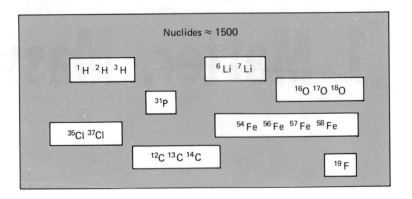

Figure 1.1 Isotopes as sub-sets of nuclides.

has an approximate isotopic composition of 75% ^{35}Cl and 25% ^{37}Cl. Its relative atomic mass is given by

$$A_r(Cl) = (75\% \times 35) + (25\% \times 37) = 35.5$$

A_r is a multiple of the atomic mass unit (amu) where one atom of the carbon isotope ^{12}C is taken as having a mass of exactly 12.000 amu.

The **Relative Molecular Mass** (M_r) is the mass of a molecule on the same scale, where one atom of ^{12}C has 12.000 amu. In practice, M_r is the sum of the A_r values for all the atoms in the molecule. With ionic compounds, for example NaCl, the term **Relative Formula Mass** is used; this corresponds to relative molecular mass.

The **Mole** (mol) is that amount of substance which contains a given number of specified units. These units may be, for example, H atoms, H_2 molecules, CH_3^+ ions, C–C bonds and so on, whether the units have an independent existence on their own or not. The number is calculated to be 6.022×10^{23} mol^{-1} and is called the **Avogadro constant**, L. For elements and compounds the appropriate amount of substance is the formula mass in grams.

1.3 Interpretation of Chemical Equations

An equation may be interpreted at two quite different levels: (a) the molar (empirical) and (b) the molecular (theoretical). Thus for the reaction

$$2 H_2 \quad + \quad O_2 \quad \longrightarrow \quad 2 H_2 O$$

we may say either

(a) 2 moles + 1 mole $\longrightarrow$ 2 moles

or

(b) 2 molecules + 1 molecule $\longrightarrow$ 2 molecules

Experiments at the molar level enable immediate deductions to be made at the molecular level. This is how, for example, the empirical formula (EF) of a compound may be calculated.

The **Empirical Formula** (EF) of a compound is the simplest ratio of the atoms of each element present. For example, it is found from chemical analysis that the EF of phosphorus (V) oxide is P_2O_5.

The **Molecular Formula** (MF) gives the actual numbers of atoms present in a molecule. For phosphorus (V) oxide the MF is actually P_4O_{10} (i.e. 2 × EF). Formulae of ionic compounds, such as NaCl, KBr, denote the simplest ratio of the ions present.

Avogadro's Principle tells us that, at the same temperature and pressure, equal volumes of gases contain the same number of molecules. Conversely, equal numbers of molecules occupy the same volume under the same conditions. It is found from experiment that a mole of any gas at STP has a volume of 22.4 dm^3 and a slightly larger value — 24 dm^3 — at RTP (see the following section, 1.4).

1.4 Ideal Gases

The kinetic model of an ideal (perfect) gas assumes that its molecules travel randomly in straight lines unless in collision with each other or with the walls of the container. The molecules (a) are point masses (have no volume), (b) have no attraction for each other or for the walls of the container and (c) collide elastically (no loss of energy, although it may be transferred between particles). Gases approach ideal behaviour at low pressures and high temperatures. The laws of Boyle and Charles hold for near-ideal gases only.

Kinetic theory shows that for an ideal gas

$$pV = \tfrac{1}{3}Nm\overline{c^2} = nRT$$

where $\quad p$ = gas pressure, Pa or atmospheres
$\qquad V$ = gas volume, m^3 or dm^3
$\qquad N$ = number of molecules present
$\qquad m$ = mass of 1 molecule, kg
$\qquad \overline{c^2}$ = mean square velocity, m^2 s^{-2}
$\qquad n$ = number of moles of gas = $\dfrac{\text{mass, } m}{\text{molar mass, } M}$
$\qquad R$ = the ideal gas constant,
$\qquad \quad$ 8.314 J K^{-1} mol^{-1} or 0.082 dm^3 atm K^{-1} mol^{-1}
$\qquad T$ = absolute temperature, K

For one mole of gas, $N \times m$ is the molar mass, M, so that

$$pV_{\mathrm{m}} = \tfrac{1}{3}M\overline{c^2} = RT$$

where V_{m} is the molar volume of the gas, V_{m} is about 22.4 dm^3 at STP or 24 dm^3 at RTP.

STP is **Standard Temperature and Pressure**, namely 273 K and 101 325 Pa (1 atmosphere).

RTP is **Room Temperature and Pressure**, 298 K and 1 atmosphere.

For convenience in calculation, 1 atmosphere is sometimes expressed as 1.013 kPa or 10^5 Pa. Occasionally pressures near to atmospheric are given in mm of Hg. 1 atmosphere is then 760 mm Hg.

1.5 Real Gases

At high pressures (when molecules are close together) and at low temperatures (when molecules are moving relatively slowly) gases depart from ideal behaviour as follows.

1. The gas volume must be corrected to allow for the volume of the molecules. The volume becomes $V - b$, where b is a constant for the gas concerned.
2. Intermolecular attraction makes the measured pressure too low. The correction constant, a, depends on the density of the gas and on its boundary surface area.

Both effects are inversely proportional to the volume. The corrected pressure is given by

$$p + \frac{a}{V^2}$$

The modified gas equation – called the **van der Waals' equation** – is thus

$$\left(p + \frac{a}{V^2}\right)(V - b) = RT \quad \text{(for 1 mol)}$$

The way in which real gases behave is shown by experimentally determined pV isotherms. Two are sketched in Fig. 1.2.

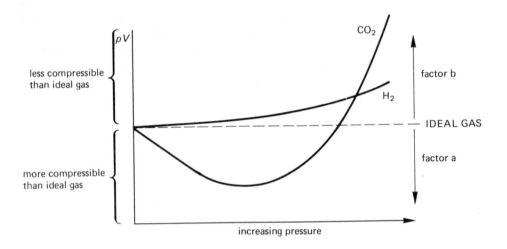

Figure 1.2 Non-ideal behaviour of gases.

1.6 Maxwell–Boltzmann Velocity Curve

At a particular instant in time, the velocities (i.e. kinetic energies) of the molecules of a gas will have various values. This distribution will change with temperature but not with time. In Fig. 1.3 the total area under each curve is constant and represents the total number of molecules in the gas.

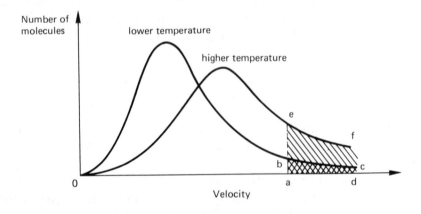

Figure 1.3 Maxwell–Boltzmann velocity curves.

The effect of temperature on **rate of reaction** may be considered as follows. Suppose that in a gaseous reaction all the molecules with a speed a or greater are able to react. The fraction reacting at the lower temperature is given by the area abcd in Fig. 1.3. At the higher temperature the larger fraction aefd can react so the rate of reaction will be greater. The speed a represents a constant, minimum amount of energy needed (k.e. $= \frac{1}{2}mv^2$) for the reaction to take place: this is the **activation energy** for the reaction. It does not vary with temperature.

1.7   Worked Examples

1.1 Finding A_r from isotopic composition

(i) Define *fully* relative atomic mass.

(ii) Calculate the relative atomic mass of natural strontium from the following data:

Mass number of Sr isotope:	84	86	87	88
% abundance:	0.56	9.9	7.0	82.6

(AEB, 1983)

Discussion and answer

(i) **See Section 1.2.**

(ii) A_r (Sr) $= \dfrac{(0.56 \times 84) + (9.9 \times 86) + (7.0 \times 87) + (82.6 \times 88)}{100}$

$= 87.8$

Check that the answer makes sense. By far the largest proportion of atoms have a mass number of 88 – the weighted average must therefore be very close to this value.

1.2 A calculation based on moles of particles

Calculate the number of electrons contained in 2.8 g of nitrogen gas. (N $= 14$, $L = 6.02 \times 10^{23}$ mol^{-1})

(SEB)

Discussion and answer

Do we know the number of electrons (e^-) in *any* quantity of nitrogen?
   Yes – in one atom there are $7e^-$ ($Z = 7$).
In one mole of atoms there are $7 \times (6.02 \times 10^{23})$ e^-.
Remember that free nitrogen is diatomic, so that:

In 1 mole of N_2 molecules there are $2 \times 7 \times 6.02 \times 10^{23}$ e^-.

Furthermore, 2.8 g of N_2 gas $= \dfrac{2.8}{28} = 0.1$ mol

$\therefore$ number of e^- in 0.1 mol of N_2 gas is

$\dfrac{(2 \times 7 \times 6.02) \times 10^{23}}{10} = 8.43 \times 10^{23}$

1.3 Reacting masses deduced from a chemical equation

In the Kroll process magnesium reacts with titanium(IV) chloride to produce titanium metal.
(i) Write an equation for the reaction.
(ii) Calculate the mass of magnesium required to react completely with 9.5 tonnes of titanium(IV) chloride.

(NISEC, slightly adapted)

Discussion and answer

(i) $TiCl_4(g) + 2Mg(s) = Ti(s) + 2MgCl_2(l)$
Since a calculation is to be based on the equation it is essential to balance it properly.

(ii) **No data on A_r values are supplied in the question: the board in question supplies a separate data sheet. For this answer we need the following data:**

Ti = 48, Cl = 35.5, Mg = 24

Substituting these A_r values in the equation of part (i) we see that

190 g of $TiCl_4$ ≡ 48 g of Mg (**molar interpretation**)
∴ 190 tonnes $TiCl_4$ ≡ 48 tonnes Mg

Divide by 2:

95 tonnes $TiCl_4$ ≡ 24 tonnes Mg
∴ 9.5 tonnes $TiCl_4$ ≡ 2.4 tonnes Mg

1.4 Calculation of an empirical formula

A compound P contains carbon 59.4%, hydrogen 10.9%, nitrogen 13.9% and oxygen 15.8% by mass. Find the empirical formula. (H = 1, C = 12, N = 14, O = 16). (L)

Discussion and answer

Since the percentages are given by mass, this is equivalent to saying that, if we had 100 g of the compound,

59.4 g would be C
10.9 g would be H
13.9 g would be N
15.8 g would be O
100.0 g total

> Check that the total comes to 100 ∓ 0.1

We first turn these masses into numbers of moles and use the fact that the *molar* ratio of the elements is the same as the ratio of the *atoms* of the elements. So we have:

moles of C = $\dfrac{59.4}{12}$ = 4.95

moles of H = 10.90

moles of N = $\dfrac{13.9}{14}$ = 0.993

moles of O = $\dfrac{15.8}{16}$ = 0.988

> Always work to at least 2 decimal places

We now divide each result by the smallest value (0.988):

moles of C = $\dfrac{4.95}{0.988}$ = 5.01

moles of H = $\dfrac{10.9}{0.988}$ = 11.03

6

moles of N = $\dfrac{0.993}{0.988}$ = 1.01

moles of O = 1.00

The empirical formula is $C_5H_{11}NO$.

Note: A difficulty with EF calculations is to know when and how to 'round off' to get whole numbers. A useful rule of thumb is that, when a molar ratio is within ∓ 0.05 of a whole number it is safe to round off. In the example we have just worked through, the final step of dividing each ratio by 0.988 might not have been considered necessary since 4.95, 0.993 and 0.988 are all within 0.05 of whole numbers. It would, however, be somewhat risky to round off 10.90 to 11.00 without careful consideration of the other ratios.

1.5 Calculation of a molecular formula from the results of combustion analysis

Spectroscopic techniques suggest that an organic compound W has a relative molecular mass of 74.0 . . . W contains only the elements C, H and O. On complete combustion, 0.0444 g of W gave 0.0792 g of carbon dioxide and 0.0324 g of water. Use the above data to find the molecular formula of W. It is essential that you explain each step of your working clearly. (H = 1, C = 12, O = 16)
(SUJB)

Discussion and answer

First summarise the data to see the problem clearly and if you prefer to work without fractions, multiply throughout by a suitable factor (e.g. 10^4).

$$W + O_2 \longrightarrow CO_2 + H_2O$$
444 g   792 g   324 g

The equation cannot be balanced because some of the oxygen is present in W and the remainder is supplied from elsewhere.

792 g of CO_2 = 792/44 mol of CO_2 = 18 mol CO_2

This will contain 18 mol of C atoms = 18×12 = 216 g of C

Similarly, 324 g of H_2O = 324/18 = 18 mol H_2O

This contains 36 mol of H atoms = 36 g of H

∴ Mass of (C + H) present in 444 g of W = 216 + 36
                                      = 252 g

∴ Mass of O present in 444 g of W = 444 − 252 = 192 g

The data can now be used to find the EF in the usual manner.

moles of C = 18
moles of H = 36
moles of O = 192/16 = 12

The simplest ratio of moles is thus C = 3 : H = 6 : O = 2
The EF is $C_3H_6O_2$.

This has a relative molecular mass of 74 amu and corresponds to the molecular formula.

1.6 A calculation using Avogadro's Principle

When 10 cm³ of the gaseous ether CH_3OCH_3 were exploded with excess oxygen, there was an overall contraction of x cm³. A further contraction of y cm³ took place when excess aqueous sodium hydroxide was added. (All measurements of gas volumes were made at 25°C and one atmosphere pressure.) Deduce the values of x and y and show clearly your reasoning.
(OLE)

Discussion and answer

First write a balanced equation assuming complete oxidation to carbon dioxide and water:

$$CH_3OCH_3 + 3O_2 = 2CO_2 + 3H_2O$$

From Avogadro's Principle (**Section 1.3**): 1 vol + 3 vol   2 vol + 3 vol

           $10\ cm^3$ + $30\ cm^3$ $20\ cm^3$ + $30\ cm^3$

The volume relationship would be correct above 100°C when water exists as a vapour: however, at 25°C the water may be considered to have entirely condensed to a liquid of negligible volume. So the first contraction $x\ cm^3$, is from $40\ cm^3$ to $20\ cm^3$. The second contraction, $y\ cm^3$, takes place because all the CO_2 is absorbed by the sodium hydroxide (making a solution of sodium carbonate). So the value of y also is $20\ cm^3$.

1.7 Finding MF of a gas from its density

A fluoride of sulphur has an empirical formula SF_4. 0.100 g of the gaseous fluoride occupies a volume of $22.10\ cm^3$ at 20°C and 102.1 kPa. Find the molecular formula of the gas. (Molar volume of a gas = $22.4\ dm^3$ at STP, standard pressure = 101.3 kPa, S = 32, F = 19).

(AEB, 1982)

Discussion and answer

Using the ideal gas law:

$$p_1V_1 = nRT_1 \text{ (for the conditions as given)}$$
$$\therefore 102.1 \times 22.10 = nR \times 293$$

Similarly, to find the volume, V, at STP

$$p_2V_2 = nRT_2$$
$$\therefore 101.3 \times V = nR \times 273$$

It follows that:

$$\frac{102.1 \times 22.10}{101.3 \times V} = \frac{293}{273}$$

$$\therefore \qquad V = 20.8\ cm^3$$

i.e. the volume of the gas at STP is $20.8\ cm^3$.

The final step involves a simple proportionality:

0.100 g has a volume of $20.8\ cm^3$ at STP
M g has a volume of $22\,400\ cm^3$ at STP
(where M g is the mass of 1 mole of gas)

$$\frac{0.100}{M} = \frac{20.8}{22\,400}$$

$$\therefore \quad M = 107.7\ g$$

$$M_r = 107.7$$

Now, the formula mass of SF_4 = 108 amu
SF_4 is therefore the molecular formula.

1.8 The behaviour of ideal and real gases

(a) Discuss briefly, but critically, *three* basic postulates underlying the kinetic theory of gases and explain the use of $\overline{c^2}$, the mean square velocity, in the equation

$$pV = \tfrac{1}{3}mN\overline{c^2}$$

where N molecules of ideal (perfect) gas, each of mass m, occupy a volume V at a pressure p.

(b) By sketching graphs of (i) pV plotted against p, (ii) p against V, for an ideal gas and a real gas (of your own choice) over a suitable temperature range, relate what is seen to the earlier discussion.

(c) Calculate (i) in terms of $R/\text{J mol}^{-1}\ \text{K}^{-1}$, the universal gas constant, the kinetic energy/J mol^{-1} of the molecules in one mole of ideal gas at 27°C; (ii) in terms of $R/\text{J mol}^{-1}\ \text{K}^{-1}$ the *root* mean square velocity/m s^{-1} of sulphur dioxide molecules at 27°C in the gaseous phase, assuming ideal behaviour; (iii) the ratio of the *root* mean square velocities of methane and sulphur dioxide assuming ideal behaviour, at 27°C. (H = 1, C = 12, O = 16, S = 32).

(SUJB)

Discussion and answer

(a) *Three postulates underlying the kinetic theory of gases*

(1) The molecules of a gas have negligible volume.

This is the case when there are relatively few molecules in the container (i.e. at low pressure). As the pressure is increased (reducing the volume of the container is one way) the space taken up by the molecules becomes an increasingly larger fraction of the volume of the gas.

(2) The molecules of a gas do not attract one another nor are they attracted to the walls of the container.

If this were so, changes of state such as the condensation of a vapour, could not take place. Intermolecular attraction becomes more important when molecules are close together (at high pressure) and when they slow down (kinetic energy reduced, as at low temperature).

(3) Molecules of a gas collide elastically.

This means that no energy is lost in collisions. If energy were lost, a gas would spontaneously fall in temperature. This does not seem to happen, so the postulate appears to be true for real gases.

In a sample of gas with large numbers of molecules, at any given instant there will be on average as many molecules moving in one direction with a mean velocity $+c$ as there are in the opposite direction with a mean velocity $-c$. The overall mean velocity will thus be zero. This applies to any direction chosen within the gas sample.

However, if all the velocities are *squared* first, all the squares will be positive. c^2 is the mean of the squares of the velocities. The square root of this value – the Root Mean Square (RMS value) – is close to the average speed of the molecules of the gas.

(b) *Departure from ideal behaviour shown by CO_2*

(i) Sketch of pV against p

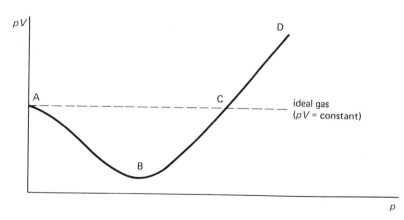

For the part of the curve at lower pressures, ABC, CO_2 is more compressible than an ideal gas. This is because intermolecular attractions assist the applied pressure. The CD part of the curve shows that at higher pressures CO_2 is less compressible than an ideal gas; the volume of the molecules is not a negligible fraction of the volume of the gas.

(ii) pV isotherms for CO_2

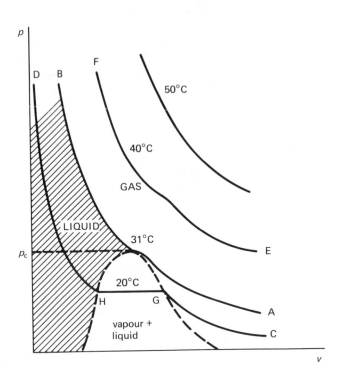

The diagram will convey the essential ideas here, with brief added comments. However, a rather full answer is given because this topic often gives difficulty to students.

AB, CD, EF are isotherms for CO_2 at the temperatures shown. Look at CD first. Beginning at point C, a mass of gas at $20°C$ is compressed. The volume diminishes along the curve CG. At G there is no change in pressure, but a dramatic reduction in volume to H — the gas is condensing due to intermolecular attraction. The curve HD shows a further small change in volume as pressure is applied to *liquid* CO_2.

Now consider the $40°C$ isotherm EF. This is more like an ideal pV curve, obeying Boyle's law. No condensation occurs here, however great the applied pressure, because at this temperature the molecules have a mean kinetic energy high enough to overcome intermolecular attraction.

AB is the *Critical Isotherm*. $31°C$ is the *Critical Temperature*, T_c. It is the highest temperature at which condensation can take place by applying pressure alone. P_c is the minimum pressure needed for condensation to occur at $31°C$. It is the *Critical Pressure*.

(c) (i) *The kinetic energy in 1 mole of an ideal gas at 27°C*

$pV = \frac{1}{3}mN\overline{c^2}$

This may be re-written as

$$2/3 \times \underbrace{1/2m\overline{c^2}}_{\text{mean KE}} \underbrace{N}_{} = RT$$

total KE

Therefore at 300 K ($27°C$),

$2/3 \times$ total KE $= R \times 300$
$\therefore$ total KE $= 450\,R$

10